# LIVING WITH ALZHEIMER'S DISEASE

## - ONE COUPLE'S JOURNEY

Frances Siegel

LIVING WITH ALZHEIMER'S DISEASE
– ONE COUPLE'S JOURNEY

2nd Edition, Published by Lulu.com

(in association with Yellow Hat Publishing, Yellow Hat Productions, Inc., 5190 Neil Road #430, Reno, NV 89502)

Illustrations by Paul Jeffrey Davids

ISBN 978-1-300-70121-7

# Living with Alzheimer's Disease

## One Couple's Journey

By Frances Siegel

I dedicate this book to my two children, who loved their father dearly and who, now sixteen years after their father's passing, still feel the great sorrow of losing him to Alzheimer's disease.

# Quotes About This Book

"*Living With Alzheimer's Disease – One Couple's Journey* poignantly portrays the joys and sorrows encountered on the journey through this incredible disease. There is a jewel for every reader among these pages – and many readers will identify closely with their own personal experiences. This book will bring comfort to many, concerns to some and inspiration to all."

**Anne Towne**, Executive Director, Greater Washington Chapter of the Alzheimer's Association

"I've read *Living with Alzheimer's Disease – One Couple's Journey* and found it to be beautifully written. The manuscript is well organized and superbly documents the progression of Alzheimer's disease and its effects on the patient and caregiver. I was very much moved as I followed the story from its onset to its inevitable conclusion."

**Daniel C. Helmstadter**, President, Scholarly Resources, Inc.

# Foreword

By Cynthia D. Steele, R.N., MPH.,
Assistant Professor Johns Hopkins School of Medicine
Co-Director Neuropsychiatric and Memory Group, Johns Hopkins Hospital

Alzheimer's disease (A.D.) is the most common cause of dementia in late life. At the time I am writing this Foreword, in 1999, this progressive brain disorder currently affects four million Americans, and that number is expected to increase to at least nine million early in the new millennium. A.D. results in a relentless course of decline for those who suffer from it. The earliest symptoms are difficulty in recalling recent events such as who has called on the telephone. Mid-stage symptoms include difficulty in finding words one wants to say and eventually the inability to understand what is being said to them. Often the next symptom is impairment in the person's ability to do things, like dress, bathe, write checks, and drive a car. Judgment and problem solving eventually erode. The last symptom to arise is agnosia or difficulty in recognizing the world around them. Thus familiar people are strangers and patients feel unsettled in their own homes, as they are no longer recognizable to them. Such patients live in an unfamiliar world with persons speaking a language they cannot understand. Even seemingly simple tasks like taking a shower become daunting, frustrating and frightening for many patients.

Eventually patients are completely dependent and must be supervised constantly.

While research has made advances in understanding the disease process in the brain and in devising treatments for the past 10-15 years, deterioration is currently unavoidable for most patients. Most patients live in the community and are cared for by their family members for as long as possible. Such constant care exacts a toll from caregivers. Even the most effective medication available today cannot substitute for the devoted efforts of the family.

Many researchers have studied the caregiving journey. Impact of these arduous tasks results in fatigue, frustration, depression and sleep disorders for many caregivers. What are not described often are the benefits that accrue from the process. One might wonder why most families do this difficult job for persons who no longer recognize them and need total care. What many will tell you is what is described so eloquently in this book. They will tell you that to care for this person is a feeling of reciprocity, that is they were kind to them and deserve kindness, that to do this job is consistent with their values and beliefs and that in the process of learning to be a caregiver, new skills are learned. Most of all they will tell you that part of what the person was in his or her nature – behavior or intellect— remain even in a very debilitated patient. Simple reactions, such as a smile of recognition when approached by their caregiver, can be very meaningful. In some ways it signifies to the caregiver that the person they know and loved is still there.

While the disease progresses relentlessly and caregiving takes its toll, there are in fact no generic patients or families. Each patient and family unit is unique and brings to the process the history of their lives together and their own individual strengths and limitations. It is only by recognizing these unique characteristics that are brought to bear on the illness that truly effective and compassionate care can be provided.

This book describes the life of a patient and those who loved him throughout the illness. It is honest in recounting the frustrations encountered but also in describing how humor was also evident and helpful. The caregiver you will read about is one who used all possible resources to help herself and her beloved husband. She approached each level of decline with sadness but with determination to continue on. It was indeed an honor to have known this patient and his caregiver and to have journeyed on with them. Caregivers will learn much from this work, as will professionals. This text chronicles not only the devastation of the disease but also the coping skills learned — and most importantly the enduring love that sustained them both throughout the illness.

# PREFACE

Alzheimer's disease (A.D.) is a devastating period of anguish for both the caregiver and the patient. The here and now is the only reality during the illness, not what was or what will be. Accepting this is very painful. It took me a long time to make peace within myself and to "live one day at a time."

I learned that A.D. has a grab-bag of symptoms that differs from person to person. Each person's experience is unique. The one constant, however, is the continuous physical and mental degeneration of the A.D. victim, eventually leading to survival only with around-the-clock care.

My husband, Jay (names of family members and the Alzheimer's disease victim have been fictionalized in this otherwise entirely factual account), was diagnosed as having A.D. in 1984, about 12-and-a-half years before his death. I was his sole caregiver until his last two years, when he was in a nursing home. In addition to Jay's aberrant behavior, I had to adjust to the loss of things we did together, and yet still attempt to keep a level of quality in Jay's life. As much as possible, I continued the activities we enjoyed. For many years during his illness I took him to museums, holding his arm for security. We went to movies, fairs, farms for apples, and to visit family and friends. We went to our grandchildrens' school activities, and they and our children came to see us.

Jay still had enjoyment in his life, despite the Alzheimer's, watching, laughing, clapping, hugging and being hugged. He would say "wonderful, wonderful," when he saw our daughter and her family walking toward him.

As the disease progressed, I adjusted his activities to continue giving him this better quality of life at his own level. Unfortunately Jay's incontinence ended many of our public activities during his last year at home. I couldn't take him everywhere any more.

So many factors enter into each person's experience with Alzheimer's disease: the closeness of the patient and the caregiver through the "normal" years, the nature of relationships (i.e., husband and wife, or parent and child), their personalities, the ability to adjust to changes throughout life, an understanding of the disease, and the possible constraints of one's financial situation.

To better understand my own situation, it helps to know our background. Jay, a Ph.D., was a professor of national renown and a prolific author. He taught at an outstanding university. I am an ex-school teacher. To help each other in our work, I proofread for him, and he set up worksheets for me. He helped at home while I took advancement courses at night.

During the many years together we always helped each other, and throughout Jay's illness, he had the same sweet, loving, helpful disposition he had had all his life. Our relationship as husband and wife and closest friends never changed. Jay never became aggressive, violent and demanding, as many Alzheimer's patients do. He never wandered, always wanting to stay close to me. This made my caregiving much easier.

The following deals with the development of the disease in my late husband, our handling of the problems, the effectiveness of research and drugs, support systems, my feelings and insights, and the roles of our two adult children.

As I recount my experiences and the lessons I learned, I hope my record can be a help to others living with Alzheimer's disease as a patient or caregiver.

— Frances Siegel

# Table of Contents

# 1. Thought Patterns of the Caregiver

In this section, I will consider all the various factors I noted in the preface as they apply to my husband and me. My thoughts and feelings followed these four patterns:

**Denial**
**Irritation and Frustration**
**Acceptance**
**Giving only love with caregiving**

1.  Denial of the diagnosis: In the early years, I always looked for ways in which my husband's symptoms were not typical of Alzheimer's disease.

2.  Becoming upset and irritated with Jay as his aberrant behavior became frequent and more intense. In this degenerative disease, it only gets worse. I experienced his three deaths: his mind, his body, and cessation of life. It was unbelievably painful.

3.  Accepting Jay's actions as an illness not controllable. This was made more difficult because his earlier years were so physically healthy.

4.  Eventually I felt within me only sorrow, love, and warmth toward Jay during his last six to seven years.

## 2. The Development of Alzheimer's Disease

For purposes of organization, I am arbitrarily dividing the years of Jay's illness into four periods: 1982-1986, 1986-1990, 1990-1994 and 1994-1996.

The first period was one that included Jay's progressive loss of abilities, confusion, and very gradual forgetfulness, which resulted in the diagnosis of Alzheimer's disease in 1984. In this period, I tried to re-teach skills he was losing, such as handling a checkbook and the math involved. In trying to make his abilities "normal again," I was denying what seemed to be happening to him. I tried to give only help and love. I attributed the changes to his overworking. But I was concerned enough to take him to a neurologist. After the diagnosis in 1984, I felt sick at heart whenever I was forced to acknowledge the diagnosis of Alzheimer's disease and what that meant for our future.

The Second Period (1986 - 1990) was when I found Jay's behavior the most difficult to live with, and when I was most frustrated and irritated. I couldn't understand why he couldn't do simple tasks. I was losing my husband and best friend. During this period of about 3 years, in order to stay in equilibrium, to avoid being angry or irritated, I wrote down incidents of aberrant behavior as they occurred, to get annoyances out of my system and not show my feelings to Jay. I wasn't 100 percent successful. I attended support groups for about seven years

(1988 - 1994). It helped greatly to be with others who really understood what I was enduring. We shared feelings and the ways in which we managed. We gave and received suggestions. We were a team and shared information about medicines, experiments, research and getting respite. No matter how much warmth and empathy others give you, only those with the experience of living with Alzheimer's disease and caregiving are really on the same wavelength about the illness. We heard about reputations of group homes and nursing homes. We gave and received suggestions on how we might handle behaviors. Now as I write this, in 1998, four years after I stopped attending support groups for home caregivers (and two year's after Jay's passing), I still meet with many of my friends from the groups. The social worker leaders gave us insights. We also benefited greatly from the assistance of the Alzheimer's Association. Reading books on Alzheimer's disease was very helpful. I got a great deal out of reading Coping with Alzheimer's by Rose Oliver and Frances Bock.

The third period (1990 - 1994) was one of gradual acceptance that the aberrations were an illness, not deliberate actions, and I realized that the deterioration and its resulting changes were not controllable by Jay or by me. Only my perception was reversible.

It became easier in these years to give only love, to touch and hug Jay continually. The irritation was completely gone by 1991. I finally knew without thinking about it that all the problems were caused by the sickness of Alzheimer's disease, and we both needed closeness and love.

The Fourth Stage (1992 - 1994) was his stay in a nursing home, where his physical illness, fevers and jaundice increased. I could no longer take care of him because his bouts of illness became more frequent and he needed medical care more readily available. Also the years of stress, emotional and physical, had taken their toll on me. I couldn't continue to care for Jay at home. I had earlier searched out and checked five nursing homes and finally selected one, where I placed him after his last hospital stay. It was terribly painful for me to place him in a nursing home. But the selection turned out to be a good one. With all of his difficult needs handled by a warm, caring and very good nursing staff, I could just enjoy being with him and loving him. I visited him six days a week at the nursing home. Jay and I went walking, lunched out, visited parks, shopped in a nearby grocery store, stopped for frozen yogurt and sang to a musical tape of show tunes I recorded for him for a small tape cassette. He had loved music all his life. I took him to musicals, some films, and parties held in the nursing home.

Each Saturday, our daughter and her family came to take him with them. He loved their visits. His face lit up when he saw them coming and he clapped his hands joyfully. (Our son and his family lived across the country in California and saw Jay much less frequently).

I attended the bi-monthly nursing home support groups. They were warm and beneficial. We had an excellent social worker group leader who handled many of our concerns in the nursing home and out. I was still a caregiver. Every nursing home patient should have a strong advocate no matter how good the home. The nursing home staff couldn't know Jay's needs as well as I did.

## 3. We Are Faced with Alzheimer's Disease

I first saw evidence of a problem in Jay that might be more than the absorptions of "an absent-minded professor" when he was about 62 or 63 years old. At the time he was under a good deal of pressure writing an "autobiography" for a famous person, consulting with the subject, doing research and oral histories for the work, advising students, mentoring dissertations, writing letters of recommendation, attending and chairing committee meetings and teaching. I had good cause to believe Jay's problems stemmed from overload. I felt "a quart bottle can only hold a quart." This went on for about a year-and-a-half.

It was at the end of 1982 while we were driving to the birthday party of his aged "autobiography" subject that I felt he was experiencing more than just burnout. Going to the party, he didn't know a route we used to take frequently from our former home. He seemed lost.

For the first time I felt he was experiencing more than "normal" forgetfulness. Within six to nine months, I realized he had difficulties of a different nature. He couldn't balance the checkbook or even figure the correct amount of the check. He had trouble with other mathematical problems, i.e., subtracting with regrouping (borrowing). He had trouble with spatial relationships, e.g., in front of, in back of. Next he began to tell time incorrectly, and then later couldn't figure out the minute hand of the clock. He didn't know or bother to see if he was getting the right change in a store, although he could usually pay the correct amount at a cash register. He was missing the concept that using a checking account is the same as getting and spending his own money. Time relationships began to go as well.

I remember that he mentioned a certain teacher we both had in college. He wondered what she was teaching now. She was in her late 50s when she taught us. She'd have been 100 at that time, if living.

When we had trouble getting a new table out of a back seat of the car, he wanted to open the trunk to make it easier. When attaching an extension cord, he attached one end to the other instead of plugging it into the wall and the appliance. These changes were occurring when I had Jay diagnosed by a neurologist who was recommended by our family doctor in 1984. I was present for most of his tests.

I learned that Jay couldn't draw a clock, count backwards by sevens, or draw a simple geometric figure, all part of the cognitive test. He had a CAT-scan of his brain, a spinal tap, an EEG, and a blood test. In addition to finding atrophy abnormalities in Jay's brain, the researchers also looked for symptoms of other dementia diseases besides A.D. Only an autopsy, however, can give an absolutely definitive diagnosis. So Jay's official diagnosis was Alzheimer's disease, but I didn't believe the diagnosis. I thought it would prove incorrect. Jay in his usual quiet, sweet way, accepted the findings and just said, "I hope I won't become a burden."

In 1984 I took him to the psychiatry and neurology departments of Johns Hopkins, a visit arranged by our HMO neurologist at my request. Johns Hopkins confirmed the diagnosis.

That began a wonderful relationship with the staff at Johns Hopkins, which lasted until Jay's death in 1996. Jay was tested mentally and physically every six months. Mrs. Steele, R.N., M.P.H., was co-director of the neuro-psychiatric and memory group (see the Foreword). She treated him herself and directed Jay's care with others until his death, even visiting him for two years in the nursing home between 1994 and 1996. I was interviewed, advised and supported by Johns Hopkins at the same time, and my husband was a subject in their experimental drug and research projects weekly (and later monthly, for years).

After Jay's death in 1996, the diagnosis of Alzheimer's disease was confirmed by an autopsy at Johns Hopkins Neuropathology Laboratory.

Included in this account is a list of research drugs and their effects, if any. They gave him MRIs and PET scans. I dedicated myself to this help for Jay and he adjusted very well with my support in everything. At this time (1984-1985) I covered for him by supplying needed words in conversations. (I often felt that people were thinking, "What an overbearing wife." But I nonetheless clarified the ideas he tried to express, and did everything I could to sustain him.)

His superior intelligence, erudition and excellent typed lectures on every topic, which formerly he scarcely found necessary, carried him through the last one-and-a-half years of teaching until his retirement in 1986.

I had been pressuring him to retire for a year. His evaluations from students were dropping from all "excellents." Two students even mentioned his forgetfulness while lecturing. I wanted him to stop teaching while he was still respected and still contributing. At this time, our children, Peter, in California, and Anne, in New England, were in complete denial of Jay's illness.

It was rare for me to have disagreements with Anne, but one evening, not long after Jay's diagnosis in 1984, I was telling Anne about some of Jay's aberrant behavior. She made various excuses for him, saying that he was just being an absent-minded professor simply getting worse with age. She refused to believe he could have Alzheimer's. I was frustrated and got angry. I said I'd never discuss Jay's problems with her again. But just the opposite turned out to be true. She was my mainstay during Jay's illness as she came to accept the diagnosis and realize the severity of his confusion.

She later moved her family from Connecticut to Virginia to be near us and give us support. It took her understanding and supportive husband more than a year to find satisfactory work in our area so they could make the move. After their relocation, she became my foundation for help and understanding, as well as my confidante. Whenever I was sick or needed help with Jay, she took him home with her for a week at a time, or whatever was necessary for my aid and respite at the time.

Peter had an even longer period of denial and often acted as though he didn't want to discuss Jay's problems with me. I was upset and asked Anne to let him know what was going on with Jay. Peter acted as though I was complaining unjustly, although actually he was very concerned. This denial later changed dramatically. Many times Peter took care of Jay for a week or 10 days by himself at his second home in Arizona. He gave me respite and gave Jay much happiness with many experiences he didn't have at home. He took Jay on short trips — swimming, into a Jacuzzi, and played tennis with him. He even took Jay horseback riding (at a slow pace.) He helped Jay with showering, dressing and his other daily needs.

Jay with Peter and Peter's two children, dressed up in western clothes for this photo in Jerome, Arizona. The photo was taken when Jay visited Peter in the southwest, a few years before Jay went into the nursing home.

## 4. Alzheimer's Behavior

During this early period of denial of the diagnosis, realization that something was wrong, and my attempts to re-teach Jay, I still believed he could control his actions. I lost patience as his behavior worsened. In order to get the frequent irritability and infrequent anger out of my system, keep my annoyance from Jay, and have an outlet, I began to keep a record of some of Jay's aberrant behavior. Writing down behaviors that upset me kept me from dwelling on them and helped me handle the situation more appropriately.

The following is my record (as originally written) of some of the gradually degenerating behavior that I wrote down during this second period (1986 - 1989). Later (1990 - 1991), I still wrote down some examples of aberrant behavior, but annoyance no longer existed. Keeping a record helped give me some security at this time.

**Behaviors:**

**END OF 1985 – 1986**

• Confusion in labeling and counting pages in his writing: 198, 199, 1000 (not 200).

- Put all coins in a drawer – couldn't buy with coins, only bills.

- Filled the car from one gas tank and checked the cost on tank across the way.

- Gas tank pull-up – always stopped the car too soon with the gas hose unable to reach the car's tank.

- When changing a parking sticker from 1985 to 1986, removed the wrong sticker from the car.

- Scraped someone's car in a parking lot but was unaware he did it.

- Going to a professional meeting out of town, he left his one-half share of our trip money in the drawer after we had divided the total between us for carrying purposes. This time I did blow my top – I didn't accept the difficulty or comfort him. On the trip, we used our charge card for everything.

- Jay had trouble getting oriented to finding and living in a hotel room. He succeeded only after five practices.

## Back Home:

• At my suggestion, he Xeroxed his attendance sheet blanks for the year. He put a date on the original, so all the sheets had the same date for the year.

• When walking with me, he always stayed one step behind me, never alongside me. I believe he used my being ahead as a cue for directions.

• Closed the trunk of the car after each package as a clerk loaded the car.

• Warmed up the car when it had snow on it. Started the car and let it stand, but didn't turn on the heater to defrost it.

• Ruined contact wire of flashlight by pulling it out when putting batteries in.

• Broke stapler putting in staples,

• Had trouble recording grades opposite the correct names in his grade book. I did it for him during his last term before retirement in 1986.

• I told him I had doctor's appointment at 4:30. If he wanted to come with me, he should be home by 3:45. Jay called at 3 p.m. to see how I made out at the doctor.

## On Vacation:

On any vacation, I had to take all the responsibility for planning, packing for two, getting tickets, and arranging transportation to the airport. I had to take Jay to the bathroom at the airport and on the plane, stand near the door until he came out, and also keep tabs on his whereabouts at all times. This made for fear and uncertainty and resulted later in my stopping all trips unless someone else was with us.

We went to La Jolla, Calif., by car after visiting Peter and Peter's family in Los Angeles.

• In the car, Jay wasn't sure where we were going, although we had talked about going to La Jolla many times.

• Disoriented in our vacation room, Jay slept in underwear instead of his usual pajamas. I didn't bother to have him change.

• Dried himself with a floor mat which was also terry cloth, instead of the towel.

• Motel key was a card with holes. He had to be told every time he opened the door to leave the key in while using the card.

•   All week he had trouble attaching the seat belt of the rental car, because it pulled from the shoulder, not from the floor as in our own car.

## Back Home Again:

•   Jay helped empty the freezer section of the laundry room refrigerator. We put food temporarily into the laundry tub for the hour it took to defrost the refrigerator. While I was upstairs, Jay ran the washing machine for his swim trunks while the food was still in the tub. The water ran out of the washer into the tub, causing a flood in the laundry room. I had to throw out the soaked food. I was so upset and angry. I thought, "I can't stay a step ahead of him."

•   Now and then he extended paragraphs instead of indenting them when writing.

•   He had trouble with appointments. We bought a small calendar notebook to keep all appointments straight to replace his old one, which was full. He ruined it by taking notes on students' papers in it. This was his last class before retirement.

•   He wrote an address: New York, New York, N.Y.

- He no longer could arrange chess pieces. He had been a very good chess player since boyhood. For me — a painful loss — what a change!

- Didn't know what "Independence Day" was, although he knew it was a celebration (1986).

- He brought home roll books from 1980 to 1985 to look up the record he had of someone in 1979.

- At the time of retirement in 1986, Jay had problems with school duties. Just after retirement, he told a student he'd write a letter of recommendation for him and then decided not to do it because he couldn't handle it. He didn't get the needed facts from the student, such as his out-of-state home address, phone number and the courses he had taken. He couldn't mentally reconcile the fact that it was a "To whom it may concern" letter with its being a law school application. We looked at term papers and grade books. He got a school record of students' classes and grades. I wrote an outline for him, and he wrote the letter. He had to be led each step of the way with much frustration for both of us. This was 180 degrees from his former efficient management of everything.

- Jay took the letter of recommendation to school to Xerox it, so if there was any trouble with the student's getting it, he'd have a copy. He forgot and mailed it without first making a copy.

## 1987 ABERRANT BEHAVIOR

Jay decided to sweep the garage when the temperature was 39 degrees outside. He didn't wear a sweater or coat in the open garage. It is hard to watch over him continually.

• He couldn't use the telephone directory. He couldn't find the "C" names.

• He couldn't distinguish between the "toast" and the "bake" settings on the toaster oven.

• He couldn't put soap in the dishwasher after using it for years.

• He couldn't start the dishwasher.

• He could no longer put the bike rack on the car. After I put it on, he seemed very confused about getting into the car.

• He had difficulty knowing some dirty clothes had to be washed. He saw them and he folded them.

• While still driving, he couldn't use directional signals correctly.

• We had two phones, but he didn't answer the one closest to him and often missed the call.

• I once asked him to put two tablespoons of powdered iced tea into a pitcher. He took two tablespoons out of the drawer and put them on the table.

• Coming back from California, while walking down the airplane aisle, he said, "Let's take these aisle seats." He had lost the concept of reserved seats.

• We had dinner on New Year's Eve at the home of some friends. On the way home, Jay asked what the occasion was. We had all just toasted each other and said, "Happy New Year."

• A student called to ask if Jay still had a copy of his term paper from three years earlier. Jay arranged to look for it for him. When the student called back, Jay had looked up his grade book instead. I said, "Just tell him you can't find it anymore." He didn't know what was wanted of him and I didn't know where to look.

## The End of Jay's Driving

Jay sometimes had trouble following road cues while driving. Once when making a left turn he ended up on the wrong side of the median strip. We managed to cut across to the correct side of the road quickly, but I knew then that I had to end his driving, and that it would be traumatic for him and difficult for me. I handled it obliquely, not head on, until I was always driving. I'd say to him "The road is being torn up and I know the detour" or "The car needs fixing" or "The tire is going flat." I thought up many excuses and sold the second car. I talked it over with a doctor at Johns Hopkins, who felt Jay should stop immediately and told him so. That was very helpful. I could be more forceful then. Gradually Jay stopped mentioning driving.

Some caregivers ask doctors for a written prescription to forbid driving. This is often the optimum method. My support group members said this problem was terribly hard to handle, because it meets with so much resistance and is such a blow to the patient and his or her independence. Some of the methods suggested by the group to stop the patient's driving are:

- Make the person's keys unavailable.

- Do it with excuses and diversions (do something else with the person) as I did.

• If the patient has a car, park it away from the house so it can't be seen or used, or sell it.

• Ask a doctor to say the sick person must stop driving. It carries much more weight with the patient than the caregiver does - especially if the doctor puts it in a prescription form.

## BEHAVIORS OF 1988 – 1989

• Since Jay continually took pictures, and I didn't like to accumulate more than a few samples of an event, it was Jay's job to put them in an album. When we came home with a large set of pictures from Peter's house (people, house, view from roof deck, trips, etc.), Jay couldn't manage the pages and plastic sheets of the album. He had questions about every picture. This is one time I exploded and felt miserable afterwards.

• I color-coded and labeled boxes for Jay's sweaters. He mixed them up anyway. I gave up quickly. I also colored and picture-coded cabinets and drawers. It didn't help.

• Jay puts a sweater on without a shirt under it now and then. When a new loss of skill begins, it gets worse with time, so I'm always upset at a new slippage of skills.

• He didn't know where the trunk of the car is.

- I shampooed the rug and put strip sheets of paper down to walk on while the rug was wet. Jay thought that he should keep off the paper and tried to walk only on the rug. I couldn't change his mind. (Insight: I can't change anything because his ability to function depends on old habits, not on anything newly learned.)

- Jay went into NIH's hospital for four weeks for hormonal studies. He didn't seem to be using his clean towels in the hospital, yet he showers every day. He must have used his roommate's towels. He couldn't distinguish between them.

- Home from NIH for the weekend, he had trouble using his old shaver. His newer one was in the hospital. The one at home isn't rechargeable. It has to be plugged in.

- Jay saw the empty trash can at 6 p.m. across the street, and wondered whether trash was about to be collected. (It has been collected at 8 a.m, for years.) It was the beginning of his losing the concept of time.

- Jay asked several times each day how much time he still had at NIH.

- Within three hours, Jay hunted for his watch five times because it was on his right wrist instead of on his left. A shunt was on his left.

• Jay was home from NIH on Feb. 24, 1988. He seemed more alert to me: He found things in the refrigerator, smiled at a funny Art Buchwald column, laughed at a subtle situation on the TV show, "Cheers." He initiated reasonable suggestions. But it didn't last more than a few days.

• He was very confused and left his swim bag in the men's locker. When I sent him back for it, he had trouble finding it. I asked the lifeguard for help. He found it. I bought a lock with a key for him. He couldn't manage a combination lock. We kept the key on a band on his wrist.

• We kept his locker unlocked soon after. He couldn't handle the key, either.

• He put on a sweater without a shirt, then put on two sweaters when told he had forgotten to put on a shirt.

• I had him practicing writing his name and date every day, so he wouldn't lose that skill.

• By 1988, he couldn't add money in his wallet anymore.

• Jay still couldn't work the button to unlock the car door on the inside in our five-month-old car.

- Jay couldn't read the parking meter times on a meter for me. I wasn't tall enough to see the hours of allowed parking. He read everything except the one fact I needed.

- Every time we went bike riding, Jay took the pedometer, which someone bought him for distances in walking. I laughed every time he did it. Many things he did at this time made me laugh.

- Jay took audio tapes out of the two case holders. Even though the cases were labeled, he couldn't put the cassettes back in the correct case.

- It is amazing the way I gradually adjust to and accept things that used to upset me. All these years he checked the doors at bedtime. For a while, I reminded him to do it and then found it easier to do it myself. I try to keep Jay in touch with everything he can do, but unwisely I often wait too long to take over!

- In order to get a salad bowl that was in the dishwasher, Jay put away the bowl and all the other dirty dishes.

- When taking about eight ounces of soda from a 16-ounce bottle, he threw the remaining eight ounces away.

- An incident outdoors upset me greatly. The whole neighborhood had the trees sprayed for gypsy moths. Jay was with me when I paid for it, and we talked about it. The airplane came between 6 and 8 a.m., back and forth over the house. Jay went out for the paper about 6:30 a.m. I was in bed and didn't pay attention and was dozing. I went out to find him about 7.15 a.m. He was standing at the open garage door watching the plane. I didn't want the fumes in the house, in addition to wanting to keep him safe, but the house door and the garage door were open. I don't know how long he was exposed to the spray. I can't anticipate and prevent everything!

- Jay began to put checks and bills in envelopes for me as I wrote checks (as a part of staying with skills of daily living). At this time, he was unable to continue doing this, so I just did it myself. It probably was another example of my keeping him at something too long and my becoming frustrated.

- The day after a CAT scan, he was more alert. We were out walking, and he said correctly that we had walked about one-and-a-half miles, after looking at his pedometer. He bought metro tickets in the library without messing up what he wanted. Do x-rays (CAT scan) help?

- If Jay bagged the newspapers for pickup, he usually included that day's paper before I'd had a chance to read it.

- By May of 1988, Jay had completely lost the ability to find the day or date on the calendar. I decided then not to try to re-teach that skill.

- After we took the car out of a parking garage one day, we passed the garage on the way home in our car. Jay said, pointing out the car window to the garage, "That's where our car is." I had to laugh.

- Jay started mixing up his clothing drawers and his clothing. He sometimes put on two undershirts. I began to lay clothes out daily.

- Jay couldn't distinguish between instant coffee and coffee for brewing — he once used drip ground coffee in his cup of instant coffee.

- I usually kept an umbrella in the car. One day we were caught in the rain without one because he had put it in the house. I handed an umbrella to him later to put in the car when we got home. He hung it up again in the closet - I put it in the car.

- Jay would say he took papers out for the trash when he meant he took in that day's paper.

- He went into the back door of the car instead of his usual passenger front door. I began to let him stay wherever he sat.

- Jay couldn't get a towel. He no longer knew where the linen closet was.

- He kept coming to me to tell me McNeil - Lehrer was on TV after I had turned it on. I guess he wanted me to watch it with him, but I was busy with dishes. I stopped what I was doing to be with him.

- He fixated on buying groceries we'd just bought.

- Instead of unplugging the vacuum for me, Jay unplugged the refrigerator.

**Deeper Confusions:**

As Jay was becoming more confused at the end of 1988, I was becoming more accepting of his actions and less frustrated. During 1989 I didn't find as great a need to unburden my annoyances on paper. I still wrote some upsetting actions down, but much less often.

I was gradually losing my expectations of Jay's abilities and just became more "there for him." I was a slow learner in that respect, but I was assuming all tasks. This reversal wasn't complete until 1990, as I gradually saw all these actions as illness.

There was a struggle in me all along between not letting him lose abilities (so he'd be a whole independent person for himself and for me) and taking the easier route of doing everything myself.

Here are listed some of the 1989 deeper confusions I did write down:

- He used my toothbrush as well as his.

- He put on someone else's coat that was hanging up in a restaurant.

- He didn't know what to do with clothes after taking them off.

- Once his underwear was a little soiled when he took it off. He hadn't cleaned himself well enough. It upset him so much that he could not dress himself after showering. I dressed him.

- He didn't know what to do with salad dressing even when a salad was in front of him.

- He began to wear odd colored socks or no socks.

- When watching TV, he would tell me there's nothing on at every commercial, even though the program wasn't over.

- He would have his coat on but kept looking for a coat to put on.

- When there was snow on the walk, he couldn't figure out where he was supposed to walk. When shoveling snow, he shoveled the grass instead of the walk, and I had to stay out with him.

- He wiped his rubbers on a rug instead of the mat.

- When he came in with his coat on, he wouldn't take it off. I ignored it. Five minutes later I would hand him a hanger for his coat and he would hang it up. Waiting a little while avoided conflict.

- He tried shoveling the driveway one day. I showed him how to throw snow on the grass. Instead, he kept throwing it in front of himself on the driveway. I gave up trying to show him what to do with the snow. He made it a much harder job.

- There were many long words he had trouble reading at this time. He had trouble with handwriting as well.

- His ability to use the house key became a hit or miss situation.

# END OF 1989 - JANUARY 1990

In Hospital for Prostate Surgery (Trans Urethral Prostatectomy)

• Incapable of pressing the button to get a nurse.

• Incapable of saying what's wrong, what he needs.

• Pulled out his IV and had to be constantly watched. As a result he developed an infection. He had to be treated with intravenous antibiotics. He kept disconnecting tubes.

Jay had to be bathed, fed, and watched for bleeding, so I hired three, eight-hour nursing aides a day to cover the 24 hours. Anne or I filled in if they couldn't get there for all the time needed. Anne was there for Jay frequently handling the problems. Since Medicare doesn't cover the expense of Alzheimer's disease, it was very expensive for us. Hospital nurses didn't have the time for the extra care the disease requires. He had to be constantly watched so he didn't remove intravenous tubes. Also, they had trouble understanding that he couldn't follow instructions. I printed a big poster, which I taped over his bed to let the staff know what his limitations were.

When we came home we continued with the aides. For the first time Jay was negative and uncooperative: he wouldn't rest, wouldn't eat, became restless, didn't want the aides, wouldn't let me take his temperature. Each instance of his saying "no" had to be handled indirectly. I guided the aides the best I could. It was so traumatic he began to lose his reading ability.

When he was better, I started him into day care two days a week. It was about five weeks after the doctor's last check-up appointment.

## COPY OF MY POSTER FOR HOSPITAL STAFF

**AUGUST 1992**

<u>HOSPITAL STAFF: PLEASE NOTE</u>:

# Handicapped

Alzheimer's

Jay _____ has Alzheimer's Disease and can't request anything. He can't use call bell.

He is continent at home, but must be shown bathroom when he is restless and can't find it.

His bowel schedule is either before going to sleep (8PM -10PM) or early morning.

He can feed himself but needs guidance.

He sometimes can't swallow pills. Please check.

He doesn't understand questions, so his answers are not to be relied upon. He may answer "great" or "yes" to any questions.

He is completely cooperative at home.

## 5. Day Care

The first day care center was the wrong one for Jay. He went there for about two months. The staff handled him badly. After his prostate operation, he was still restless and kept going to the bathroom unnecessarily. I had told them about his operation and that the doctor said he probably had after-effects of the operation and was insecure, and that it would go away. They suggested I withdraw him and bring him back in a month. I didn't mind withdrawing him, because I thought the place was poorly run: They spoke down to him, gave him childish coloring to do, and didn't want the aides to oversee his many short bathroom visits even though it was a temporary condition. He removed himself at day care and just sat in a comfortable chair. He was at day care because he had Alzheimer's disease, yet they didn't tolerate non-conformity.

The social worker at my support group told me about a good day care center where she was a part-time staff member. What a difference this made! It was excellent.

I took Jay's records and enrolled him at the Fellowship House for two days a week. The place was well enclosed, so he was free to move as he liked, even into a fenced-in yard. The staff was warm and encouraging. He took part in walks, music and some games. He watched others a good deal. They took him and others who were able to go to theater events and out to lunch. They had regular staff and volunteers. The social worker tested Jay's comprehension. The nurse checked him if needed. Jay made many friends, as did I. I joined the weekly support group there. Gradually, I increased his attendance to three, then four days per week. It was a lifesaver for my physical and mental health.

## MID 1992:

His next hospital stay for skin melanoma of the leg caused continence to deteriorate, because he wasn't allowed to walk freely.

Everything happens gradually. Once again, I turned to Johns Hopkins for advice. The Johns Hopkins staff told me to keep him walking, at least once a day, or he'd forget how to walk. I did this. Again, we had around-the-clock nursing aides for about nine days in the hospital and home. Putting myself on his bathroom schedule - every two-and-a-half hours - resulted in few mishaps. As he recovered, he became completely continent again at home and at day care until late 1993.

After the melanoma operation, I started him back at the day care center. I took him in at 9:30 a.m. and the bus brought him home at 4:30 p.m. At this time, he spoke very little. He had echolalia. He echoed the last part of people's speech - usually about four words. One day, we all had a good laugh at the day care center.

There was a Japanese man at the day care who previously had spoken English, however, with his Alzheimer's disease, he had reverted to Japanese only. He and Jay usually sat side by side. The man spoke Japanese to Jay, and Jay echoed back the last four words he heard. They got along very well holding this conversation, both happy with the response of the other. They continually had the staff in stitches. On their daily walks, Jay and another friend held hands; Jay was the guide and leader. This man, like Jay, was a gentle, sweet man. His wife and I are still good friends, even now that both men have died (within two months of each other).

During Jay's last year of day care, he began to have bouts of illness, fever, nausea and jaundice. These bouts usually lasted six days. The doctors found that he had some blockage.

The doctor told us that an operation might help him, or else that he might become totally unable to function after such an operation. We (the doctors, our children and I) decided against surgery. We treated the bouts of illness with bed stay and antibiotics. He had a good quality of life, and we wanted to keep it that way for as long as possible. His third attack and hospital stay of eight days resulted in his admission to a nursing home, where he lived for two-and-a-quarter years.

# 6. The Nursing Home

Jay was taken from the hospital Aug. 4, 1994, as he was recovering from his bout of fever, nausea and jaundice. The hospital wanted me to take him home after eight days. He was mainly bedridden and was having trouble walking. He had a stiff gait when he did walk. He needed physical support and complete care. I had him taken by ambulance to a nursing home in Rockville, Maryland, which I had selected six months earlier, in case I'd ever need one. I was very lucky that they had a vacancy and that they could take him immediately. At first I thought the nursing home stay would be temporary. His blockage symptoms subsided, and for about two weeks, he was getting his physical therapy for walking. His speech was very minimal: just short phrases and some repetition of words and echolalia. His understanding was very limited, but hand indications helped him understand. If I said, "Let's sit down," and indicated the chair, he knew what I meant. At this time I had many sleepless nights debating permanent nursing home placement for Jay. During that first month's stay, I made up a list of pros and cons, so I could keep reading them to help me decide. The nursing home social worker said I didn't have a one-month time limit, I could decide any time I liked. If I wanted I could have him readmitted after taking him home.

## ADVANTAGES OF NURSING HOME PERMANENT PLACEMENT

• I don't have to worry about care for Jay's recurrent jaundice and getting Jay to a doctor. The nurses at the home give care and contact a doctor when Jay is sick.

• I don't have to keep Jay on his two-and-a-half hour bathroom schedule or clean him after incontinence. They do it.

• It is much easier physically not having to get Jay on and off the toilet (with my health problems, that is not an easy task).

• When Jay is sick, I don't have to find good help in the house to give me relief.

• I don't have to arrange Jay's short stays at a group home (which I did three times last year, so I could go out of town).

• With my health problems getting worse, I don't have to worry about Jay's care when I'm not well. I usually had to call Anne to come get him. What would I do without Anne's caring for Jay!

• At home I have this serious concern: What if something should happen to me? Jay's needs would not be cared for (food, toilet, sleep in bed), and he would rattle around the house until someone discovered us. He would probably sit in the car in the garage as he sometimes does when he wants me. Friends call to keep tabs on us daily, but that only helps minimally. With Jay in a home I don't have this heavy burden.

• I have freedom of movement and feel more relaxed with Jay in the nursing home.

• I don't have the physical strain of cleaning him and the linens when he throws up, and the strain of getting him to the doctor.

• Now that he's in a nursing home I don't feel so burdened and insecure. I just have strong feelings of love for him all the time. I'm so happy when I'm with him or take him out the six days a week I visit him.

## DISADVANTAGES OF NURSING HOME

• I feel despondent at home without Jay, even given the fact that there's no communication when he's around.

• When he's away I miss Jay terribly, and I don't feel he's completely mine, after 53 years of our being part of one another.

• I feel his care isn't as good as when I'm doing it, even though the medical nursing care is better than what I can give.

• At the nursing home, he's deprived of my awareness of his every need, mood, happiness and unhappiness, time for active movement and change of activity. I respond instinctively.

• I miss his laugh, his hand clapping, hugging and being hugged, and sitting down to help him with meals. He fulfills my need to be needed.

• Jay loves to be with me. He strokes me, puts his head on mine, always holds my hand and sometimes can and does say "I love you." I hate depriving him of his home with me.

• I don't know if he is aware of the sickness around him and how it makes him feel in that environment. I believe deep within his brain he is more aware than I can realize. Every now and then he shows "the window of light" when for a few seconds he makes a completely lucid statement or shows some inner hidden knowledge.

• He seems to show that he's upset when I kiss him "good-bye" - he repeats "okay, okay, okay," and he looks joyless. My saying, see you tomorrow" seems to be just words to him without meaning.

- Visiting the nursing home is depressing for me, with seeing all the handicaps and sickness.

Note: These feelings about the nursing home came from my written record during Jay's first month there (August 1994). Afterwards, I saw and knew each person there individually and didn't see the home and patients only as a depressing, sick place. Just as Jay was an individual sick person, so did the others became personalized to me. The staff too became warm, caring, helpful individuals, not just part of a nursing home. I grew to care for many of them very much. This, of course, was a gradual change that came with time. And now that Jay is no longer living, I still attend the nursing home support group meeting about once every other month, to keep up with the friends I made and still feel close to Jay in that environment.

After much weighing and soul searching, with Anne and Peter's support, I decided to make Jay's nursing home stay permanent. I was sure after two months that it was the best way.

In addition, our fine Kaiser Perrnanente primary care physician told me that he'd be coming to the nursing home every other month to check Jay medically and review his nursing records (which he did).

Jay's bouts of illness became more frequent and severe with the jaundice lasting up to seven days. During his last six months, his attacks came about once every six weeks. He received antibiotics, special diets, bed care as needed and warm attention.

Each time he was sick, I stayed all day holding his hand, putting cool compresses on his head, giving fluids and love. The nursing staff showed they cared for him and gave excellent care, checking on him frequently.

During one of his later bouts of illness, when he was very sick, one of the nurses sat with us for about three-quarters of an hour, lightly rubbing his chest and reassuring him so he'd know someone was with him. She spoke soothingly to him and spoke to me about what seemed best for him. When she had to leave, I continued this treatment as she showed me.

When Jay died from his last bout with the blockage (septicemia caused by gall bladder inflammation), two nurses asked us if they could kiss him. Others had tears as they held his hand. They found him to be the same warm, sweet, undemanding, delightful person we did, and I feel that up to his death he had a good quality and enjoyment of life.

## 7. Insights

Insights into Research:

In the early days at the time of diagnosis, Jay and I both wanted him to try experimental drugs and research, both for his sake and for that of future generations. We found that all researchers were happy to get subjects.

Everything was free of charge to us as eligible subjects. We knew that Jay tolerated drugs and medicines well, and that he wasn't an apprehensive person. We felt good about his becoming part of the research projects at Johns Hopkins. The nurses, doctors, administrators and lab operators all worked with us and for us.

FOLLOWING IS A LIST OF THE PROGRAMS HE ENROLLED IN AND THE RESULTS:

Johns Hopkins, 1984 – 1996:

Cognitive testing every six months. His test answers went from 27 out of 30 correct in 1984 to three out of 30 correct in 1992, and those three were "repeat after me" questions. With his echolalia, that was a cinch.

NIH, 1986:

This involved the use of Acetocholine, which was designed to block whatever inhibited the transmitter choline from reaching the brain. There was no improvement in Jay from this drug. Also, there was a PET Scan test in 1986. NIH researchers saw the affected parts of the brain but didn't tell me anything of significance. They said they observed atrophy.

Johns Hopkins, 1986:

Physostigmine - just two injections, a week apart – no visible difference.

Johns Hopkins, 1986:

Physostigmine followed by Pet Scan.

NIH, 1986:

PET Scan. Visits for testing Jay's abilities.

Johns Hopkins, 1987 – 1988:

Testing of the experimental drug THA. We had been reading about THA and wanted to be part of the research. Their cognitive study showed no significant changes, but Jay seemed more alert during this period.

NIH, 1/26/88 - 2/24/88:

In-patient hormonal study called SMS - no change.

Johns Hopkins, 1988:
MRI, PET Scan, followed by another MRI. These were close together and were more for research use than for Jay's diagnosis.

Johns Hopkins, 6/88:

Johns Hopkins – second THA experiment - Jay was barely eligible cognitively, but the terrific Dr. K said that since I felt he was more alert during the last THA experiment, they'd include Jay again. Unfortunately, there was no improvement.

Johns Hopkins, 1988:

Language ability test. Nuclear medicine evaluations.

Georgetown Hospital, November 1988:

MRI and a CAT Scan — both showed atrophy. The drug being tested had only a number designation. I didn't write down what it was. They saw no change in the testing.

Johns Hopkins, 1990:

Long term physostigmine testing. Jay was in clinics once a week for four months with blood tests and physicals. Then I was given a physostigmine home program for Jay. He took pills daily and was checked weekly and then monthly for side effects as well as abilities for two years. No tangible results.

Johns Hopkins, 1994:

In 1994 he was to go into an Alzheimer's Genetics Research Study, but his illness and nursing home placement precluded that.

Every time I went to Johns Hopkins, I answered questionnaires for comparisons with former times, received emotional help, was educated as to the changes, and came away less troubled because of their support. They had me call them with any questions I had. I felt as though we belonged to them.

Unfortunately, none of the research drugs seemed to help Jay. But I sometimes wonder whether Jay's very slow progression with Alzheimer's disease was due to the many drug experiments he took part in. Perhaps there was some benefit. There were fourteen-and-a-half years from the time I first sensed something was wrong until his death. Articles on Alzheimer's disease indicate that the disease generally lasts from four to twenty years. Jay certainly had a very gradual mental decline.

IN THE YEARS OF SUFFERING THROUGH JAY'S ALZHEIMER'S DISEASE, I LEARNED MANY THINGS:

- The way I viewed events could generally make them bad, tolerable, or funny. Most weren't innately so.

- Since I couldn't change my situation, ultimately I had to change the way I viewed it.

- The behavior of an Alzheimer's disease victim isn't by choice. The patient is sick. He or she should not be blamed any more than a patient who has cancer.

- Each loss of skills presents a need for new creative ways to cope (such as my handling the driving problem.)

- The struggle between having the patient continue former tasks to keep him independent and whole, and not letting frustrations overwhelm me, was a difficult balance, one I wasn't completely capable of handling. Many times I should have taken over sooner before I felt upset.

- It is important not to take apathy, hostility, lack of feeling in the patient, and negativism in the patient as a personal rejection. It is the illness.

- I said to myself, "If I were sick like him, how would I feel or cope?" "What would I need others to do to make this more bearable?"

- I was lucky in that I didn't have to handle delusions or aggression or wandering in Jay. Many Alzheimer's disease patients have a complete imbalance of their "time clock" and are up all night. I didn't have to cope with that, either. Medical aid and advice are necessary in these cases.

- Jay's total incontinence occurred about a year before nursing home care, so I only had to handle it for a short duration. A handheld shower for easy cleaning was one of the best things I ever bought. Jay liked to hold it himself, with my guidance, as we cleaned him.

- We caregivers can work to lessen our angry responses but must realize we are human beings with limitations and accept that. We're not perfect, and we try to do our best. This is a long-term chronic illness, and we know there is only a downhill direction. How can we not have unwanted reactions to all the unlimited stress?

- Lessening expectations, not expecting Jay to be what he once was, was a very gradual process. Once I reached that point after some years, I was a much less burdened caregiver and could give unconditional love.

- In Alzheimer's disease, the brain has lost the ability to see beyond the immediate. There is a sequence of mental processes in following directions or questions. The Alzheimer's disease patient can't recall, identify, or understand in order to give a normal response. Realizing this helped me be more patient with Jay.

- Humor and a sense of humor alleviate pain. When I laughed Jay laughed too. He seemed so happy. We had many incidents I laughed at. Here are two:

He finished lunch in the nursing home, pushed his chair back and started repeating "Okay, okay..." showing he was finished. The nurse came over to him and said, "Let's go, Mr. Okay." Jay looked at her with dignity and corrected her with: "Dr. Okay."

That story went around the nursing home and provided much laughter.

One day I took him to the bathroom at the nursing home, before taking him for a walk. When he was finished, we fixed his pants. I fastened his belt and off we went. We held hands, listened to a tape, sang and I talked. Suddenly he stopped, looked very upset and said "oh-oh, oh- oh, oh-oh." I looked at him and found his pants around his ankles. I laughed so hard I had trouble pulling up his pants and properly tightening his belt this time. He laughed at my laughing.

We both felt good.

• It seems unlikely, but sometimes the disease getting worse will alleviate the worst behavioral actions. The patient grows more withdrawn, less relating, loses fears. This loss of contact with the environment can result in the end of aggressions or wandering or other actions.

• Problem solving is often very difficult, but continually necessary. One incident stands out for me: One day in about 1991 we were walking at Great Falls. Jay went into the men's restroom and I waited outside. It was a cold day, and the park seemed empty. I waited and waited and finally I opened the door to the men's room and said to the only closed booth, "Jay, are you all right?" He said, "I can't come out." He didn't know how to get the locked door open. I was afraid to leave him to hunt for the park police. I put my coat on the floor of the next booth, lay on it and looked up into his booth from the small underneath space. I directed him from the floor on how to put his hand on the lock and push it open. We were successful. I knew then that I'd have to take him into the ladies' room whenever we were out and he needed the bathroom. The looks and remarks we got made this very uncomfortable. In the ladies' room, I was always saying, "He has Alzheimer's and needs help." I asked an artist friend of mine to make a card saying "Alzheimer's Handicapped," which she designed and put in a plastic pin-on holder. I always had it with me. She made 200, and I bought the pin-on plastic holders. We donated them to the Alzheimer's Association. I used it for about two years until incontinence necessitated a special undergarment. It solved one bathroom problem - awkwardness in using the ladies room.

• Putting Jay on a bathroom schedule and learning how to handle incontinence are problems I had to solve which each person learns to cope with in his own way. I didn't get upset, just found a bathroom, cleaned, and changed him. I always carried soft paper, wipe-ons, and undergarments in the car. I never had to use the clean clothes I had kept in the car.

• It is vital to be in the hands of a doctor you feel confidence in, one who can give or recommend proper help for the patient and for you. For me, Kaiser's recommending us to Johns Hopkins worked very well.

• In reacting to Jay's needs I tried to keep all words brief and to minimize explanations, and to give any true or untrue reason that would avoid confrontation. I always used diversion, too. I also learned that what works one day may not work the next.

- Starting in 1988, I saw an estate planning lawyer who specialized in senior affairs to find out what was needed. He drew up a general durable power of attorney and a durable medical power of attorney for me, since Jay wasn't capable of handling any affairs. I think this is vital! It must be done while the patient can still agree to documents and sign them - I certainly used them.

- A nursing home, no matter how good, must know that the patient has a strong advocate. I was Jay's strong advocate.

- Knowing that expectations of yourself have to change is also very important. There are many things I adjusted to not doing any more, such as having dinners for friends, because I never knew what needs Jay might have at the time. I gradually adjusted to limiting my way of life. Life couldn't be as it was.

- When I took Jay anywhere, I learned to be completely at ease when people looked at him because of his need for help, or his stiff walking gait, or his repetition of my words, or his "okay, okay, okay." I actually felt, "That's their problem, not mine," and somehow felt superior because of what they didn't understand.

- Memory loss is only one part of Alzheimer's. The patient's brain can't process what is said or done. He can't identify, have judgment, do what is needed, or utilize messages he receives. Knowing the ramifications of any action is lost for the Alzheimer's disease patient.

• The Alzheimer's Association is an excellent source of advice, references, and literature.

• It is important to accept with ease any offered support that is helpful. I had friends and family to confide in and who provided encouragement. I also had loving Anne to take over readily when I needed it. I was very fortunate.

# 8. Conclusion

Alzheimer's went on so long for us that I stopped feeling the pain of missing what used to be and just lived and dealt with the present. The only reality was helping and loving a husband with a fragmented partial mind who enjoyed much of life and loved and was loved in return. I tried to give his life quality with feelings of happiness and pleasure at each stage of his illness and to keep normalcy in mine.

I was lucky that Jay still communicated (in a very limited way) with me. Also, I was lucky that he was still walking until he died, and that his sweetness never ended. As we walked, he stroked my back, put his head on mine, and he put his arm around me. When I stopped to speak to someone, he pulled my hand. He wanted all my attention. He watched me continually, eyes following me as I walked in or out. Many caregivers aren't that fortunate. Most patients become bedridden and can no longer communicate. Some show no response at all to any stimuli. Jay and I never experienced those miserable symptoms. He was still a person.

Jay died in the nursing home on Dec. 6, 1996, with complications from fever and jaundice illness. I felt unable to handle the many necessary arrangements at his death. Anne, with help from her husband, and Peter, took care of everything. I looked on, just answering the questions they asked me.

Anne and Peter gave me endless support. They made tapes for me, framed letters and newspaper accounts about Jay and made a letter album for me of everything they knew I'd treasure. I also have an audio tape Peter made of his last visit to Jay that my son made by hiding (without my knowledge) a small tape recorder. So now I can still hear Jay's voice, singing, laughing and repeating his phrases on tape. Anne and her husband made a video tape at the university's memorial to Jay, bringing back to me our "normal life." I'm so grateful to my children to have these parts of Jay's life.

Jay and I met during my first year (his second year) of college, married while still in college and had been together through 55 years of a great marriage. I count my blessings, despite our painful years with Alzheimer's disease. I intend to stay in our home to keep our life together around me. I'm taking courses, tutoring, meeting dear family and friends, and carrying on the tasks of daily living, but nothing fills the emptiness in me left by Jay's death.

*Jay and Frances together*

# About the Author

Frances Siegel grew up in Oyster Bay, Long Island, New York. She was born in 1921 and met her husband, Jay, at Brooklyn College. After their marriage, the couple moved first to Washington, D.C. and then to Kensington, Maryland, followed by a move to Bethesda, Maryland. Frances taught fifth and sixth grade for her entire career, until her retirement.

The parents of her students acknowledged in many thank you letters their gratitude for the excellent education she gave their children and for the high self esteem she instilled in them. Throughout her life she devoted herself to her children and to Jay, and during his illness she dedicated herself to making Jay's life liveable.

Jay was a tenured professor of history at a major university throughout his forty year career. He also wrote a textbook published by Random House, which Frances helped edit, and he is noted for many major scholarly contributions in his field.

Frances' daughter Ann and her husband moved to southern California about eight years after Jay's death, and Frances sold the family home in Maryland and moved with them. Frances settled in a senior apartment in Pasadena. Currently in 2013 she is 91 years old, lives independently with part-time help, and she sees her children every week and her grandchildren frequently.